YOUR KNOWLEDGE HAS VALUE

AF300891

- We will publish your bachelor's and
  master's thesis, essays and papers

- Your own eBook and book -
  sold worldwide in all relevant shops

- Earn money with each sale

Upload your text at www.GRIN.com
and publish for free

The Origin of "Complex Tones". Multiphonic Sounds generated with Wine Glasses and Woodwind Instruments

Alexander Rehm

Bibliographic information published by the German National Library:

The German National Library lists this publication in the National Bibliography; detailed bibliographic data are available on the Internet at http://dnb.dnb.de.

ISBN: 9783346867957
This book is also available as an ebook.

On the origin of the "Complex tones" as part of Multiphonic Sounds generated with wine glasses and woodwind instruments

A.Rehm

Summary

In this study, it could be confirmed that wine glasses can generate multiphonic sounds as a result of a hit with a metal stick on the bowl of the wine glass (hit-excitation). It could be demonstrated that at least two Eigenfrequenzen (= natural frequencies) of the excited wine glasses with frequencies in Hz of "F1" and "F2" (with F2<F1) will generate complex tones with frequencies in Hz of "F2+F1" and "F2-F1" – which is a typical characteristic of multiphonic sounds known from woodwind instruments. Based on the analysis of the decay kinetics of the Eigenfrequenzen and the complex tones after a hit-excitation of the wine glass, the hypothesis has been developed that the oscillations F1 and F2 build a power-profile which can be mathematically described as a multiplication of the amplitude-values of both oscillations and that this power-profile is the driving force to generate oscillations of some areas of the walls of the bowl of the wine glasses with the frequencies "F2-F1" and "F2+F1". The fact that the phases of the oscillations of the complex tones match the phases of the oscillations forming the quasi periodic oscillation of the power-profile is an argument supporting the hypothesis. For multiphonic sounds generated with saxophones and clarinets, the same observation could be made, although one of the two complex tones shows a slight phase-shift vs. the respective oscillation of the power-profile. So, the developed hypothesis on the origin of complex tones as part of multiphonic sounds generated with wine glasses seems to be valid for woodwind instruments as well. Further investigations may be needed to confirm the hypothesis and to understand the slight differences observed in woodwind instruments concerning the phase of the oscillation of one complex tone and its respective oscillation of the power-profile.

Table of Contents

Introduction

Multiphonic sounds became an integral part of modern music, but also of musical science (Ref. 1-4). These sounds fascinate musicians, composers, audience and researchers, who want to understand and explain the physics behind this phenomenon. It has been described previously that "real multiphonic sounds" generated with woodwind instruments have a certain structure of the acoustic signals (Ref. 5, 6) containing "complex tones" and therefore differ significantly from "multiphonic sounding Harmonics" generated with string instruments like a piano or a cello which miss those complex tones (Ref. 7). In case a multiphonic sound of a woodwind instrument contains two basic signals (F1, F2) of different frequencies (Hz), the frequencies of the detectable complex tones follow the equation: $n*F1 \pm m*F2$ (n, m are integer numbers >0). A typical multiphonic sound generated with a woodwind instrument consists of the following three tone-groups: 1) basic frequency F1 (Hz) and related harmonics with frequencies of: $n*F1$ (n is an integer number >1); 2) basic frequency F2 (Hz) and related harmonics with frequencies of: $m*F2$ (m is an integer number >1) and 3) Complex tones which frequencies following the equation: $n*F1 \pm m*F2$ (Ref. 6, 7). Although strong evidence has been provided that a pair of complex tones with given frequencies of "$n*F1 + m*F2$" and "$n*F2 - m*F1$" (with $n*F2 > m*F1$) are real derivatives of the tones with frequencies $n*F1$ and $m*F2$ (Ref. 7), the concrete mechanism of this phenomenon has not been described yet. Early data describe that non-periodic oscillations of multiphonic sounds in woodwinds contain chaotic dynamics (Ref. 8) whereas Rehm et al. proposed a certain type of "Coupling of the two standing waves within the instrument" as the basic mechanism (Ref.7). Linke et al. described an "Impulse pattern formation within the instrument" as the basic principle to generate a multiphonic sound in woodwind instruments (Ref.9). Vergez et al. presented some interesting findings on quasi periodic sounds generated with simple flute-instruments (Ref. 10) and because of the simplicity of these instruments, this might help to clarify how multiphonic sounds are generated in wind-instruments. The major problem in investigating the origin of the complex tones as part of the multiphonic sound generated with woodwind instruments is the complexity of these instruments as such, combined with the complexity of the parameters influencing the sound. Fortunately, it has been reported recently that "Wine glasses" which are often used by musicians as parts of glasharps can generate sounds containing all three types of tone-groups which are characteristic for multiphonic sounds of woodwinds (Ref. 11). As wine glasses have a less complex structure than woodwind instruments, we have chosen wine glasses as "simple tube-instruments" to investigate the principle of origin of complex tones as part of multiphonic sounds. Especially detailed analysis of the kinetics and phases of the oscillations of the two basic frequencies and the complex tones as their derivatives, should help to clarify the basic mechanism behind the formation of complex tones.

Material & Methods

Recordings of sounds (wav-files) of wine glasses have been performed as described previously (Ref. 11). Analysis of wav-sound files and calculation of power-spectra (Fast Fourier Transformation= FFT) of those sounds have been done with the software Praat (Ref. 12). Praat has the function to extract the oscillations at certain frequencies via a reverse FFT from the power spectra. For this extraction method, it is crucial to set the right bandwidth in Hz in order to gain a good signal/noise ratio and a good time resolution. For the analysis presented in this study, the bandwidth has been set in most cases to 20Hz, as this setting has shown to give the best results. In some cases also bandwidth of 40Hz or even 60Hz have been used. Reducing the bandwidth below 20Hz will have the effect that fast transients of the intensity of the oscillation at the selected frequency may not be reflected correctly in the extracted wave.

Multiphonic sounds generated with a King tenor saxophone have been downloaded from Ref. 13 and other multiphonic sounds of woodwind instruments from different sources (Ref. 7, 14, 15) have been used for the analysis with Praat (Ref. 12).

The method described by Denninger (Ref. 16) to compare the intensities of the two peaks of an Eigenfrequenz (=natural frequency) in a power spectrum of a wine glass after a hit-excitation has been used to localize the points on the rim of the bowl of the wine glass with minimum or maximum relative movement during the production of an audible sound.

By using the integrated functions of Praat software, amplitude values of sine-like oscillations could be calculated from dB-values of these oscillations and vice versa. Also, the "power/sec" and the "energy" of sine-like oscillations could be calculated, as the power correlates with the 2.power-value of the amplitude – if the amplitude of an oscillation is given in Pascal (Pa), the power of this wave is given in Pa^2.

In this study, we introduce the "power-profile" of two parallel sine-like oscillations at frequencies F1 and F2 as the result of the multiplication of the amplitude-values of these oscillations at a certain time (ti). The equation of this mathematical calculation is:

Amp-value-F1(ti) * Amp-value-F2(ti) = Power-pointF1/F2(ti)

Amp-value-F1(ti) and Amp-value-F2(ti) are the amplitude-values of the waves with the frequencies F1 and F2 (Hz) at the time (ti).

Plotting the values of the "Power-pointsF1/F2" in the time domain will give the "power-profile" of the parallel oscillations with the frequencies F1 and F2 ".

Results

a) Analysis of sounds of wine glasses

Although each single wine glass is generating a specific and "individual" sound after an excitation through a hit (hit-excitation) against the bowl (due to its specific parameters like shape, diameters, thickness of the glass and its overall structure; Ref. 17) the generated sounds show common spectral characteristics. The power spectra are dominated by the Eigenfrequenzen (natural frequencies) of the wine glass, where several signals of the Eigenfrequenzen consist of a double-peak with only a few Hz difference (Ref. 11, 16-20). It has been demonstrated recently that the power spectra of sounds of wine glasses contain signals – beside the Eigenfrequenzen – which a) can be classified as harmonics of one of the Eigenfrequenzen and also signals which b) can be classified as complex tones as the frequencies of their dB-maxima are given by the equation : F-Ct(Hz) = F-Ei(Hz) $\pm$ F-Ey(Hz) – where F-Ct is the frequency of the dB-maximum of the Complex tone and F-Ei & F-Ey are the frequencies of the Eigenfrequenzen Ei and Ey in Hz (Ref. 11). It has further been described that the intensity of the Eigenfrequenzen show a typical decay kinetic resulting in a linear decline of the dB-value vs. time with a negative slope-value given in -dB/sec. For some wine glasses, the negative slope-values and the frequencies of the various Eigenfrequenzen show a linear correlation as well (Ref. 11).

A typical result of the analysis of the decay kinetics of the various signals of a sound generated by a wine glass after hit-excitation is presented in figures 1A-1F. Figure 1A shows the amplitude-values vs. time of the sound generated by a hit against the bowl of the wine glass, and Figure 1B displays the power spectrum of this sound. The Peaks within the power spectrum can be classified as a) "Eigenfrequenzen", b) "Harmonics of the Eigenfrequenzen" and as c) "Complex tones" (see Material & Methods and Ref. 11) which is demonstrated in Figure 1C. Most of the Eigenfrequenzen show the expected double-peaks (Ref. 17-20) whereas the maxima of the Eigenfrequenzen F3, F6 and F7 show a single dB-maximum only. For F6 and F7, a double peak can also be generated by changing the position of the hit against the bowl of the glass slightly (data not shown). In case of F3, this will not result in a double peak. Based on the knowledge that acoustic signals generated by the oscillation of the bowl and rim will have a double-peak as a consequence of the two possible oscillation-modes (Ref. 17) it can be concluded that the oscillation responsible for the Eigenfrequenz F3 is not generated by the bowl, but by other parts of the wine glass. It is worth to note that in all wine glasses examined (with different shapes and sizes) signals classified as complex notes could be detected. Most of the complex tones had frequencies which were a sum of the frequencies of two Eigenfrequenzen – only in a few cases also complex tones with frequencies resulting from a difference of two Eigenfrequenzen could be detected. Figure 1D displays the decay kinetic of the intensity (dB) of the Eigenfrequenzen F1 and F2, as well as of the complex tone with the frequency: F1A+F2A. As expected the decay of the

Eigenfrequenz of F1 is slower than the decay of F2, but both show a typical kinetic with a period of a linear reduction of the intensity (dB) vs. time (sec). The intensity of the complex tone F1A+F2A is significantly lower compared to the Eigenfrequenzen and shows a faster decay-kinetic. Although the determination of the decay-kinetic of the complex tones is more difficult and - due to the lower intensity of these signals - results in a lower R^2-value of the linear regression, it could be observed that the slope of the complex tone is highly similar with the sum of the slope-values of the two Eigenfrequenzen. This is demonstrated in figure 1E where the slope of the complex tone is determined to -25,2 (dB/sec) and the sum of the slope-values of F1A and F1B result in a value of -26,5 (dB/sec). The dB-values of the multiplication of the amplitude-values of the Eigenfrequenzen F1A and F2A and a further multiplication by the constant: 1.2 is shown in figure 1E. The determined slope of this multiplication of amplitudes is -26,1 (dB/sec) which is close to the expected value of -26,5 (dB/sec). In this case, the factor 1.2 was chosen in order to give similar amplitude values in the linear dB-decay phase as the complex tone. It is worth to note that the chosen multiplication-factor has no influence on the dB-decay-kinetic and on the slope-value. Data on the decay of the Eigenfrequenzen F1, F2 and F4 (F3 excluded as this is not an oscillation of the bowl) are plotted in figure 1F and confirm the already published linear correlation of the frequency and the decay-kinetic of the Eigenfrequenzen caused by the bowl (Ref. 11). Further, it is shown that the 1st Harmonic of F2A and the complex tone (F1A+F2A) have a decay kinetic which do not match the linear regression curve of the Eigenfrequenzen.

Using a different wine glass of the same type will give similar, but not identical results. The power spectrum of the sound generated through hit-excitation (figure 2A) is similar to the spectrum in figure 1B, but shows significant differences, especially in the intensities of the Eigenfrequenzen. Figure 2B shows the decay-kinetic of F2A and F4A and of the complex tone (F2A+F4A). The slope of the dB-decay of the complex tone (F2A+F4A) is -43,7 (dB/sec) which is close to the value of -44,3 (dB/sec) for the sum of the slope-values of F2A (-17 db/sec) and F4A (-27,3 dB/sec) and the value of the multiplication of the amplitudes of F2A and F4A at $-$ 44,1 (dB/sec) – see figure 2C. This glass also shows the linear correlation of the frequency and the decay-kinetic of the Eigenfrequenzen as well as the deviation from this correlation by the decay-kinetics of the 1st Harmonic of F2 and the complex-tone (F2A+F4A) - see figure 2D.

Based on the given examples (figures 1 & 2) and similar analysis using other types of wine glasses, it can be concluded that the amplitude of complex tones fulfill the following equation:

Amp-complex tone at frequency (F(i)±F(y)) = Amp-F(i) * Amp-(Fy) * E-factor-Ct(F(i)±F(y))

Short version: Amp-Ct(F(i)±F(y)) = Amp-F(i) * Amp-F(y) * E-Ct(F(i)±F(y)) with: F(i)>(Fy) in Hz

E-factor-(Ct) is a positive constant specific for the respective complex tone (at the certain frequency F(i)±F(y)) interpreted as "Efficiency-factor". The higher the value for the E-factor, the larger is the amplitude of the Complex tone vs. the result of the multiplication of the amplitudes of the two respective sounds F(i) and F(y). This could be interpreted as a higher efficiency of the sounding system to create the complex tone with the frequency "F(i)±F(y)".

It has already been proposed that a specific coupling of two standing waves of different frequency within a woodwind instrument causes the complex tones, and that this yet not understood mechanism can be described by the multiplication of the amplitude-values of the two oscillations (Ref.3). This is similar to the equation above for wine glasses. The sound of wine glasses do not result from standing waves inside the bowl, but from oscillations mainly of the bowl itself (Ref. 16, 17). So it is unlikely that the underlying mechanism to generate complex tones is a mechanism happening in the medium air but in the material glass. This means that the wine glass itself should develop an oscillation with the frequency of a complex tone F(i)±F(y) initiated by the combination of the two oscillations with the frequencies F(i) and F(y). The software developed by the University of Munich "TUM" (Ref. 17) has been used to study and simulate the oscillations of the rim and bowl of wine glasses.

Simplified projections of the first and second oscillation-modes of the bowl of a wine glass are shown in figures 3A and 3B. The left parts of the figures focus on the areas of maximum (large arrows) and minimum (small arrows) movement of the rim, whereas the right parts demonstrate the two possible types of oscillation for each oscillation-mode, which are responsible for the observed double-peaks in the power spectra. The audible sound is mainly generated by the changing air pressure within the bowl due to the changing volume of the bowl as a result of the oscillation of the rim and the walls of the bowl (Ref. 16, 17). If the 1st mode (being responsible for the signal of the first Eigenfrequenz) and the 2nd mode (responsible for the second Eigenfrequenz) oscillate in parallel with their two oscillation-types, we expect a complex oscillation pattern of the wall and rim of the bowl - see simplified projection in the left part of figure 4. The projection in figure 4 is based on an experiment where the points of the maximum and minimum movement of the rim for the two oscillation modes have been determined (see Materials & Methods). It is obvious from the right part in figure 4 that the combined oscillation of the 1st and 2nd oscillation mode generate areas of the rim which show a maximum oscillating movement especially if the maxima of both modes are closely related but also areas which will show only a minor or even no oscillating movement. These "non-oscillating-areas" (no-areas) are caused by the overlapping of the 1st and 2nd oscillation mode. So these no-areas act like a stable wall against the oscillating air-pressure generated by the combined oscillations of the 1st and 2nd mode. This means that the power generated by the oscillating air-pressure will act fully against the no-areas, but no oscillation of this area at the frequency of the 1st and 2nd Eigenfrequenz can occur. The power of an acoustic wave correlates with the 2.power of its amplitude – so if the amplitude is given in "Pa" the power of the wave will be given in "Pa^2". It can be assumed that beside the power of the oscillations of the 1st and 2nd Eigenfrequenz also a combined power of both amplitudes may act against the no-areas. This additional "power-profile" could be calculated as a multiplication of the amplitudes of both oscillations (resulting in a Pa^2-value). It has been shown that the multiplication of two sine-waves with the frequencies F1 and F2 in Hz (with F2>F1) result in an oscillation consisting of two waves with the frequencies F2-F1 and F1+F2 (Ref.7). So we may interpret the "power-profile" of the 1st and 2nd Eigenfrequenz as a power being able to initiate an oscillation of the no-areas. The no-areas will neither oscillate at the frequencies of the 1st - nor of the 2nd Eigenfrequenz, as explained above. But as the power-profile contains components with frequencies of "F2-F1" and "F1+F2", this impulse might initiate a corresponding oscillation of the no-areas. By comparing the phases of the oscillations of the complex tones with the phases of the 1st and 2nd Eigenfrequenzen and the phases of the oscillations within the power-profiles of both Eigenfrequenzen, it should be possible to find arguments for the ideas presented above. If the complex tones in wine glasses are the result of the power-profiles, the phase of the oscillation of the complex tone with the frequency F(ct), and the phase of the oscillation of the signal within the power-profile with the same frequency F(ct), should be identical.

The data presented in figures 5 and 6 for a wine glass filled with 50ml of water demonstrate the fulfillment of the above-mentioned expectation. As, it has been previously reported that wine glasses with a relatively large volume may generate sounds of higher intensity if filled with some water (Ref. 11), such a condition have been chosen for the experimental analysis. The data presented in figure 5 showing the decay kinetics of the Eigenfrequenzen F1 (633,3Hz) and F2 (1696Hz), the complex tone "F1+F2" (2329Hz) and of the "Multiplication of amplitude-values of F1 and F2", generally confirm the findings displayed in figures 1 and 2 with high precision. Figure 6A shows the oscillation of the complex tone (red curve) at the frequency of F1+F2 (2329Hz) and the "power-profile" of the Eigenfrequenzen F1 and F2 (black curve) resulting from the multiplication of the amplitude-values of both oscillations. It is obvious from figure 6A that the power-profile is not an oscillation with one single frequency, but is a combination of two oscillations of different frequency. An FFT-analysis confirms the expected overlapping of two oscillations with frequencies at F1+F2 (2329Hz) and F2-F1 (1063Hz) as parts of the power-profile. In figure 6B, the two oscillations with frequencies at 1063Hz (brown curve) and 2329Hz (blue curve) forming the power-profile are also plotted into the graph. Therefore, figure 6B clearly demonstrates that the phase of the oscillation of the complex tone is identical to the phase of the oscillation within the power-profile having the frequency of the complex tone.

In another experiment with a similar wine glass, complex tones with the frequency F1+F2 and F2-F1 could be detected, although the signal at F2-F1 was weak. Figure 7A shows the oscillations of the

Complex tones at 1063Hz (F2-F1) and 2329Hz (F1+F2). It was expected that the addition of both oscillations will result in an oscillation pattern which should be highly similar to the power-profile of the oscillations F1 and F2. This is not the case (see figure 7B) but if the phase of the oscillation at 1063Hz is inverted (which corresponds to a phase-shift of 50% of the respective wavelength) the sum of both oscillations shows a high similarity to the power-profile (figure 7C).

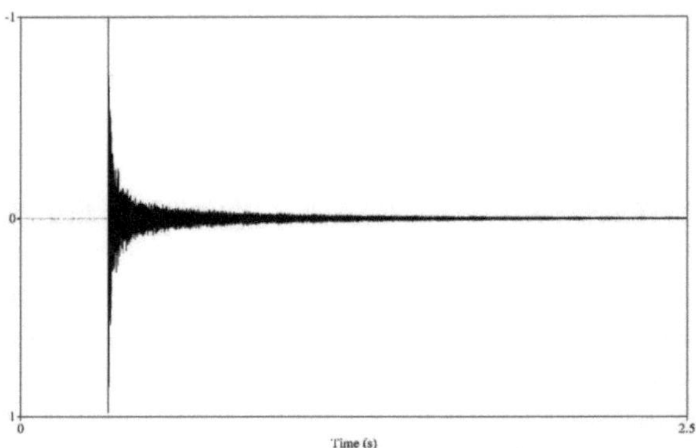

Figure 1A: Radiated sound (y-axis: amplitude rel.) of a wine glass after a hit excitation (see Materials & Methods). Time-frame: two seconds. Hit against wine glass at: t= 0,326 sec

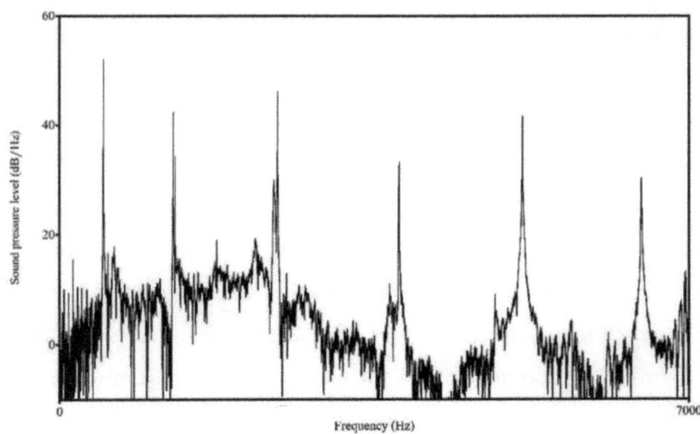

Figure 1B: Power spectrum (dB vs. frequency in Hz) of the sound displayed in figure 1A, generated through Fast Fourier Transformation (FFT – see Materials & Methods). Frequency-range: 0-7000Hz.

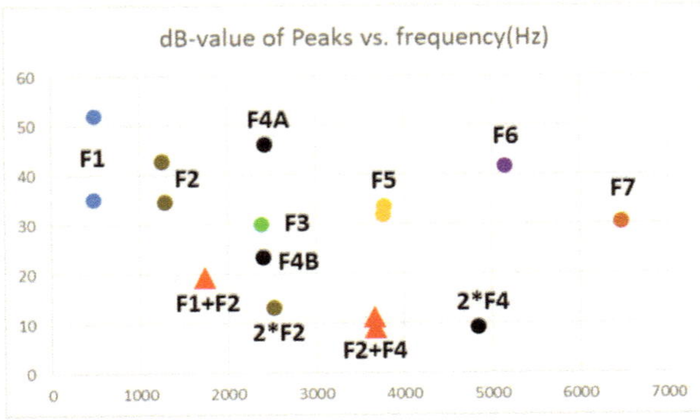

Figure 1C: dB-values (y-axis) of classified Peaks from figure 2A plotted vs. their frequency in Hz (x-axis). The seven Eigenfrequenzen are classified as F1 to F7, the 1.st Harmonics of F2 and F4 are classified as 2*F2 and 2*F4, and the Complex tones are classified as F1+F2 and F2+F4 (red triangles).

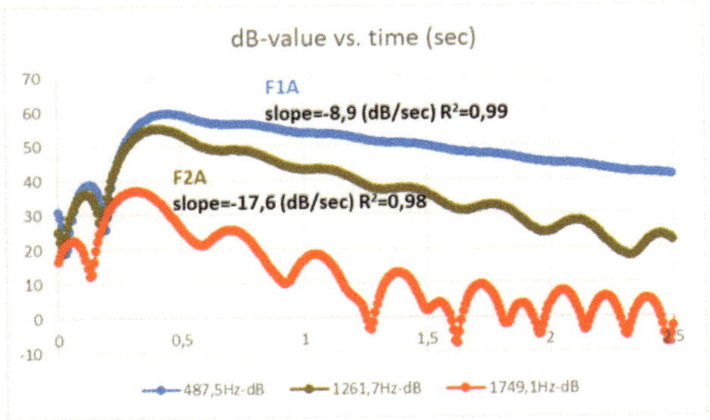

Figure 1D: dB-kinetics of the Eigenfrequenzen F1A and F2A and of the Complex tone (F1A+F2A) with a Peak-signal at 1749,1 Hz. For F1A and F2A the dB-decay is given as the "slope" during the linear phase in dB/sec which reflects the decay kinetic of the respective signal. R^2 values of the linear regression analysis to determine the slope are given.

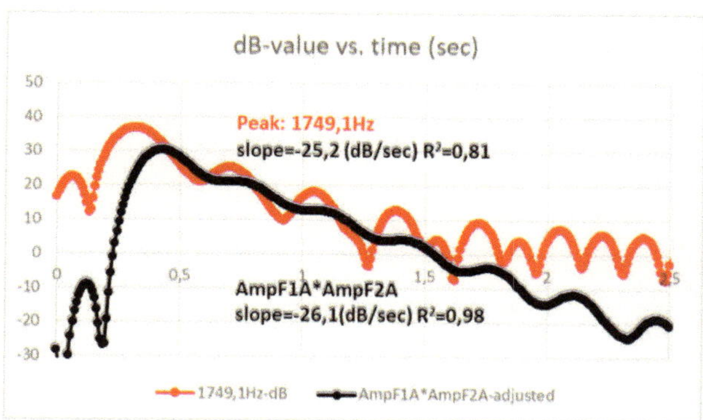

Figure 1E: dB-kinetics of the Complex tone (F1A+F2A) at 1749,1 Hz (red curve – same as in figure 1D) and the calculated and adjusted dB-values of the multiplication of the amplitudes of the oscillations of F1A and F2A (black curve). The dB-decay is given as the "slope" during the linear phase in dB/sec, which reflects the decay kinetic of the respective signal. R^2 values of the linear regression analysis to determine the slope are given.

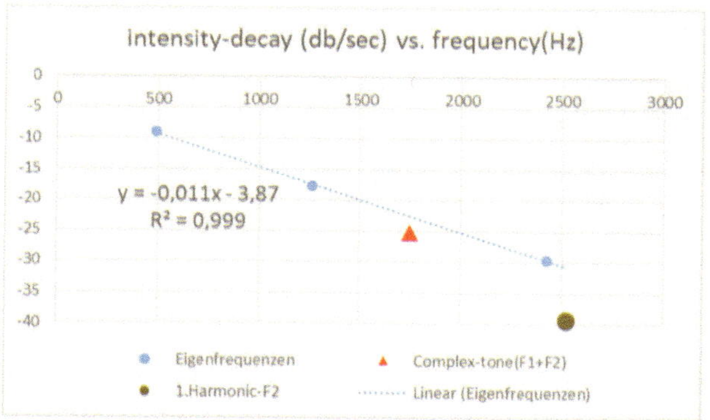

Figure 1F: Negative slopes in dB/sec as parameters for the decay kinetic of acoustic signals of the Eigenfrequenzen F1, F2 and F4 (blue dots), the 1st Harmonic of the Eigenfrequenz F2 (brown dot) and the Complex tone (F1+F2) plotted vs. the frequency of the signals. The linear regression function of the Eigenfrequenzen with the respective R^2 value is given in the figure.

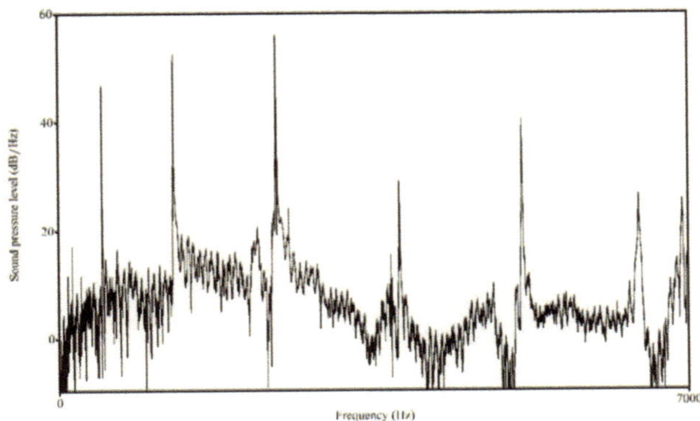

Figure 2A: Power spectrum (dB vs. frequency in Hz) of the sound after hit-excitation of a wine glass similar in type to the wine glass used in figure 1. Frequency-range: 0-7000Hz.

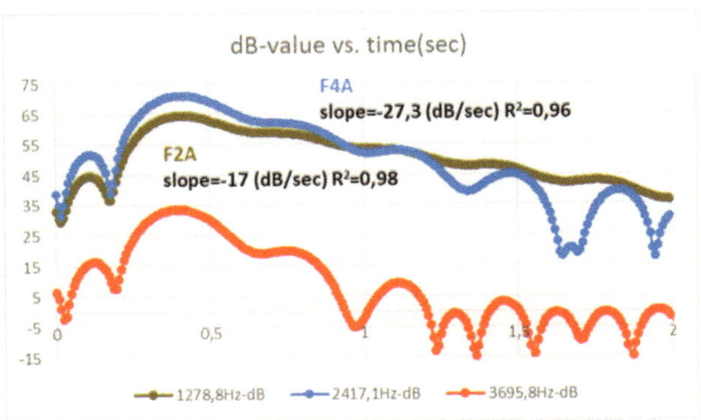

Figure 2B: dB-kinetics of the Eigenfrequenzen F2A and F4A and of the Complex tone (F2A+F4A) with a Peak-signal at 3695,8 Hz. For F2A and F4A the dB-decay is given as the "slope" during the linear phase in dB/sec which reflects the decay kinetic of the respective signal. R^2 values of the linear regression analysis to determine the slope are given.

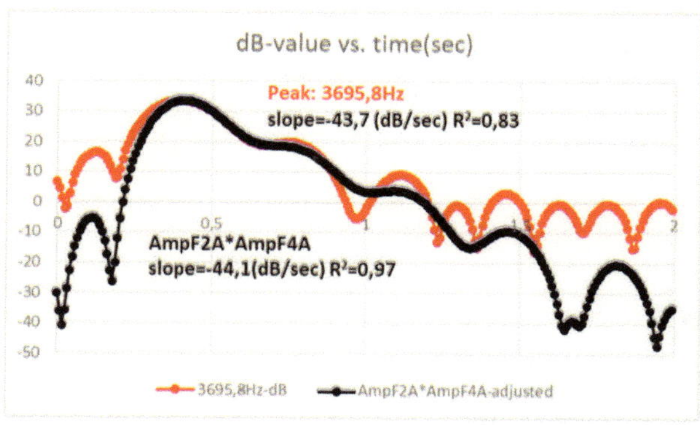

Figure 2C: dB-kinetics of the Complex tone (F2A+F4A) at 3695,8 Hz (red curve – same as in figure 2D) and the calculated and adjusted dB-values of the multiplication of the amplitudes of the oscillations of F2A and F4A (black curve). The dB-decay is given as the "slope" during the linear phase in dB/sec, which reflects the decay kinetic of the respective signal. R^2 values of the linear regression analysis to determine the slope are given.

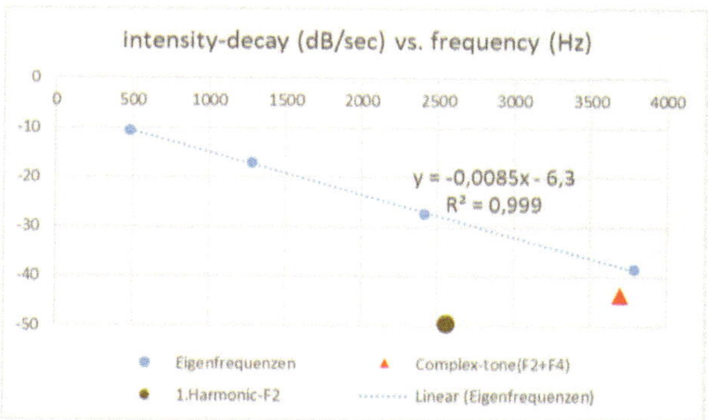

Figure 2D: Negative slopes in dB/sec as parameters for the decay kinetic of acoustic signals of the Eigenfrequenzen F1, F2, F4 and F5 (blue dots), the 1st Harmonic of the Eigenfrequenz F2 (brown dot) and the Complex tone (F2+F4) plotted vs. the frequency of the signals. The linear regression function of the Eigenfrequenzen with the respective R^2 value is given in the figure.

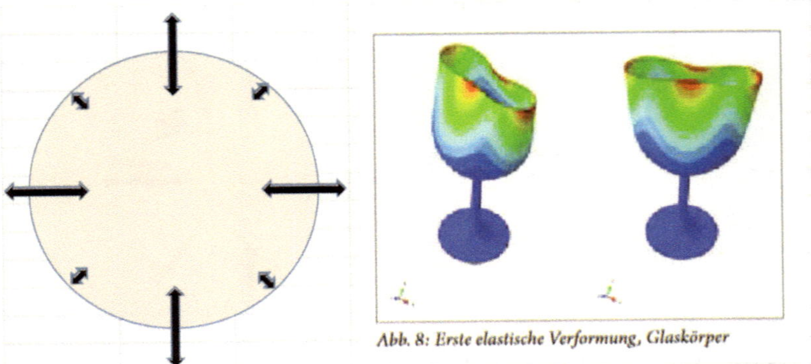

Figure 3A: Left: Simplified projection of the areas of maximum (large arrows) and minimum (small arrows) movement of the rim of the wine glass for the 1st oscillation mode. Right: 3-D projection of the two oscillation-types of the 1st oscillation mode of a wine glass –reprinted from Ref. 9.

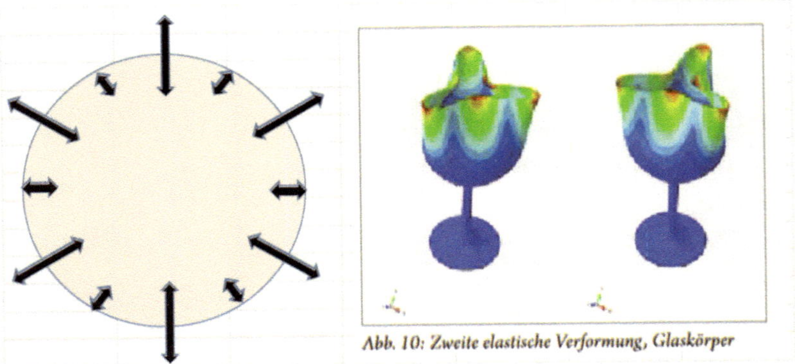

Figure 3B: Left: Simplified projection of the areas of maximum (large arrows) and minimum (small arrows) movement of the rim of the wine glass for the 2nd oscillation mode. Right: 3-D projection of the two oscillation-types of the 2nd oscillation mode of a wine glass – reprinted from Ref. 9.

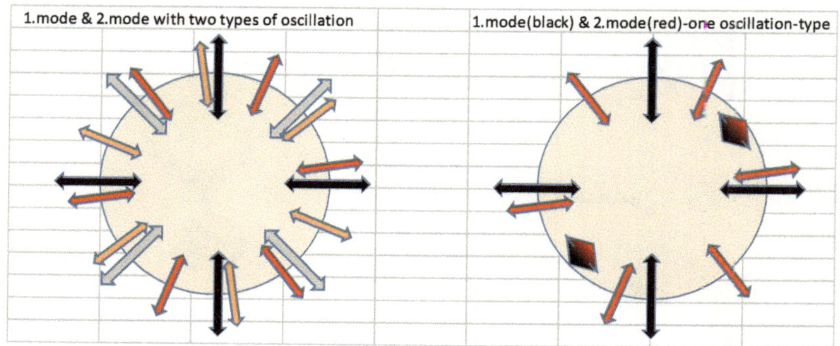

Figure 4: Left: Simplified projection of the areas of maximum movement (arrows) of the rim of the wine glass for both oscillation-types of the 1st oscillation-mode (black & gray arrows) and the 2nd oscillation mode (red & pink arrows). Right: For simplification only one oscillation-type of the 1st and 2nd oscillation mode is shown (black & red arrows) as well as the areas of the rim which do not participate in the oscillation (rhombus).

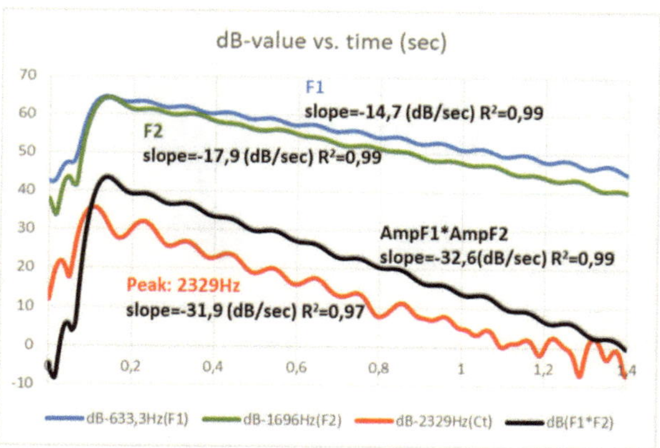

Figure 5: dB-kinetics of the Eigenfrequenzen F1 (blue) and F2 (green), of the Complex tone (F1+F2) at 2329Hz (red) and of the "multiplication of the amplitude values of F1 and F2" (black). For each curve, the dB-decay is given as the "slope-value" during the linear phase in dB/sec, which reflects the decay kinetic of the respective signal. R^2 values of the linear regression analysis to determine the slope are given.

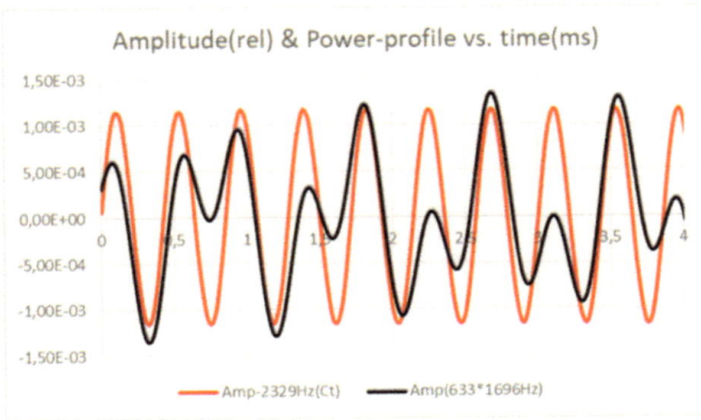

Figure 6A: Amplitude-values of the oscillation of the complex tone at 2329Hz within a time period of 4ms (red curve) and values of the "power-profile" (=multiplication of the amplitude-values) of the Eigenfrequenzen F1 at 633Hz and F2 at 1696Hz (black curve).

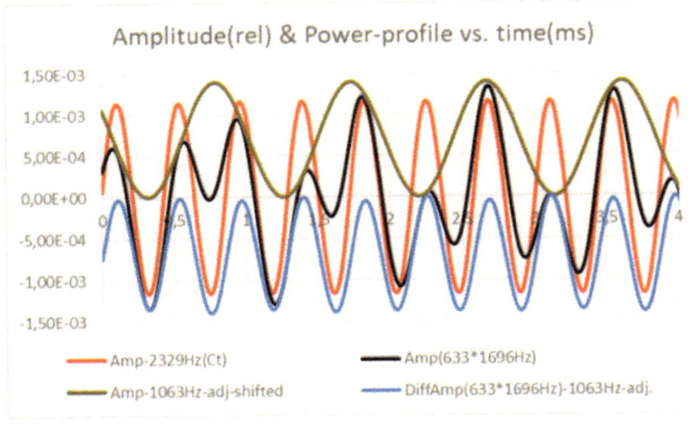

Figure 6B: Amplitude-values of the complex tone at 2329Hz (red curve) and values of the power-profile of Eigenfrequenzen F1 and F2 (black curve) from figure 6A. The brown curve shows the amplitude values of a mathematically generated oscillation with a frequency of 1063Hz – which refers to the difference of the F2 (1696Hz) and F1 (633Hz). These amplitude values are adjusted for phase and intensity in such a way that the difference of the power-profile values and the amplitude values of this oscillation results in a sine-like oscillation with a frequency of 2329Hz (blue curve).

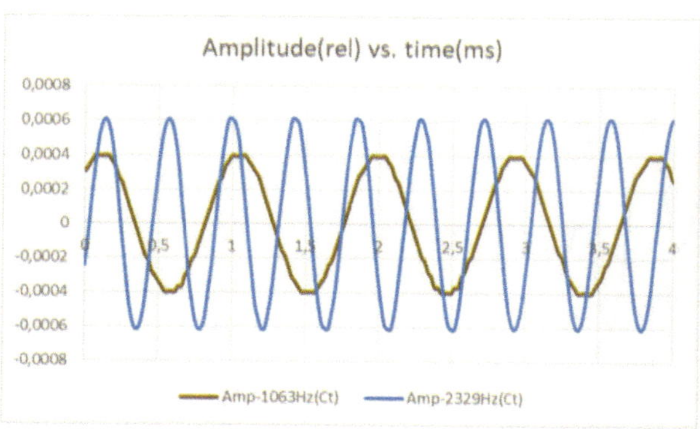

Figure 7A: Oscillation waves of the Complex tones F1+F2 (2329Hz) and F2-F1 (1063Hz) extracted from the power-spectrum of a sound generated through hit-excitation of a wine glass filled with 50ml water. (Remark: data displayed in figures 6 and 7 are generated with the same wine glass but at different experiments)

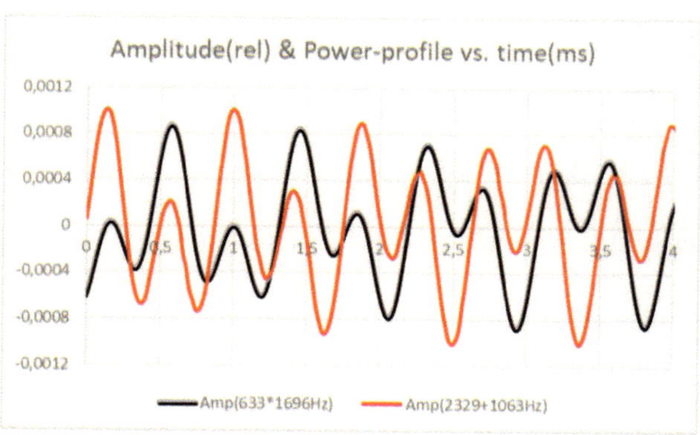

Figure 7B: Values of the power profile of the Eigenfrequenzen F1 (633Hz) and F2 (1696Hz) plotted against the time (black curve). The red curve is a result of the sum of the amplitude-values of the oscillations of the complex tones F1+F2 (2329Hz) and F2-F1 (1063Hz).

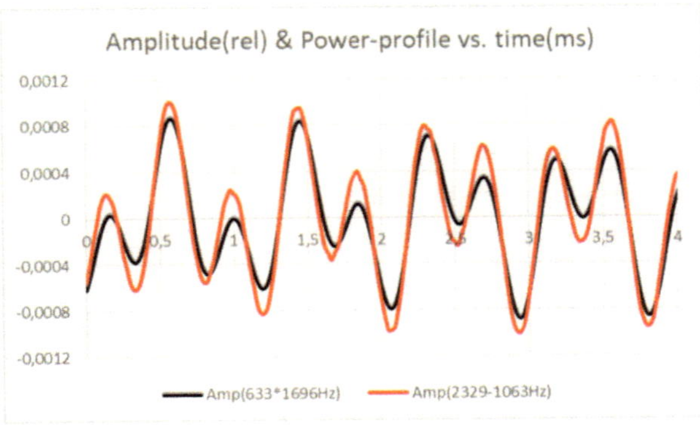

Figure 7C: Values of the power profile of the Eigenfrequenzen F1 (633Hz) and F2 (1696Hz) plotted against the time (black curve - same as in figure 7B). The red curve is a result of the sum of the amplitude-values of the oscillation of the complex tone F1+F2 (2329Hz) and the inverted amplitude-values of the complex tone F2-F1 (1063Hz).

b) Analysis of multiphonic sounds of tenor saxophones and clarinets

Figure 8A shows the decay of the oscillation of a sound generated with a saxophone (Ref. 13). In this case, the reduction of the intensity over time is a result of a controlled reduction of the air-flow through the instrument by an experienced musician. In contrast to the intensity decay of the sound of a wine glass generated through a hit-excitation where the glass characteristics determine the decay process, the player is (beside the instrument itself) a "key parameter" defining the sound characteristics. The power spectrum of this sound demonstrates its "multiphonic character" as it shows two basic frequencies at 130Hz (F1) and 415Hz (F2) and related harmonics of these two signals. Further, complex tones can be identified (see red arrows for complex tones at 546Hz, 701Hz and 962Hz in figure 8B). The amplitude-values of various extracted signals of this multiphonic sound are plotted in figure 8C and 8D. Using these data, the dB-decay kinetics of the complex tones could be calculated and compared to the dB-kinetics of the power-profiles of the respective Eigenfrequenzen. These comparisons are shown for the Complex tone at 546Hz (F1+F2) in figure 8E and for the complex tone at 962Hz (F1+2*F2) in figure 8F. Although the precision of the data is relatively low, it is obvious that the dB-kinetics of the complex tones show similarities to the corresponding power-profiles – a phenomenon which could be observed for wine glasses as well (see figures 1E, 2C, 5).

The analysis and comparison of the phases of the oscillations of the Complex tones and the oscillations forming the power-profiles of the respective basic frequencies for this sound of the saxophone are shown in figures 9 and 10. Figures 9A and 9B demonstrate that the phase of the Complex tone at 546Hz is identical with the phase of the 546Hz-oscillation, which forms one part of the power-profile of F1 (130Hz) and F2 (415Hz) - the same finding as for wine glasses (see figures 6 and 7). The amplitude-values of the oscillations of the complex tones at 701Hz (2*F2-F1) and 962Hz (2*F2+F1) extracted from the spectrum shown in figure 8B are plotted vs. a time period of 10ms in figure 10A. The different y-axis for both curves demonstrate that the amplitude of the oscillation at 962Hz is approx. 20-times larger than the amplitude of the oscillation at 701Hz (see figure 10A). According to the above proposed relevance of the power-profile of F1 and F2 as driving force for the complex-tones we should expect, that a sum of the amplitude-values of the oscillations at 701Hz and 962Hz (after an adjustment to the same maximum amplitude-values) and the curve of the power-profile should be highly similar. This is demonstrated in figure 10B, where the slight deviation of both curves are caused by a slight phase-shift of the Complex tone at 701Hz vs. the same oscillation of the power-profile. The phase-shift of the 701Hz complex-tone is 0,12ms, which represents 8% of the wavelength.

The comparable analysis of a recording of the same "multiphonic sound" as shown in figure 8, but played by a different musician on a different tenor saxophone, is presented in figure 11. It is obvious that the sum of the amplitude-values of the two complex tones at 288Hz and 546Hz (adjusted for the same maximum amplitude) does not match the power-profile of F1 (129Hz) and F2 (416Hz) – see figure 11A. This match could be achieved by shifting the phase of the complex tone F2-F1 (288Hz) by 1.066ms representing 31% of the wavelength of 288Hz, whereas the phase of the complex tone F2+F1 (546Hz) is unchanged.

The analysis of multiphonic sounds generated with a clarinet lead to highly similar results concerning the two basic frequencies and the related complex tones as those reported for saxophones. This is demonstrated in figure 12 where a clarinet multiphonic (Ref. 14) shows the same phenomenon concerning the phases of the oscillations of the power-profile of F1 and F2 and the oscillations of the related complex tones. In this case, the phase of the complex tone at 1060Hz must be shifted by 0.068ms (which represents 7% of the wavelength) to achieve a match with the power-profile.

Summarizing the analyzed data generated with wine glasses, saxophones and clarinets the following can be concluded: to achieve a match for the curves of the power-profile of F1 and F2 and the sum of the amplitude-values of the two complex tones F1+F2 and F2-F1 in the same time period a) for wine glasses either no phase shift of the complex tones is necessary or a phase shift of one of the complex tones by 50% of the wavelength (= inversion of a sine-like oscillation) and b) for woodwind instruments a phase shift of >5% and <45% of only one of the complex tones is necessary.

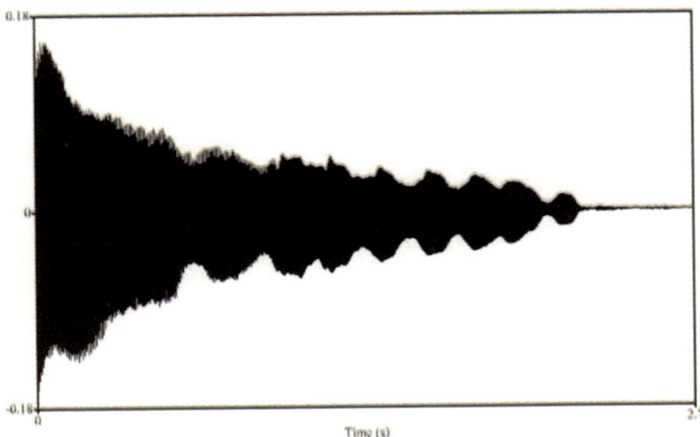

Figure 8A: Oscillation of a tenor saxophone sound (downloaded from Ref. 13) in the time frame of 2.7 seconds.

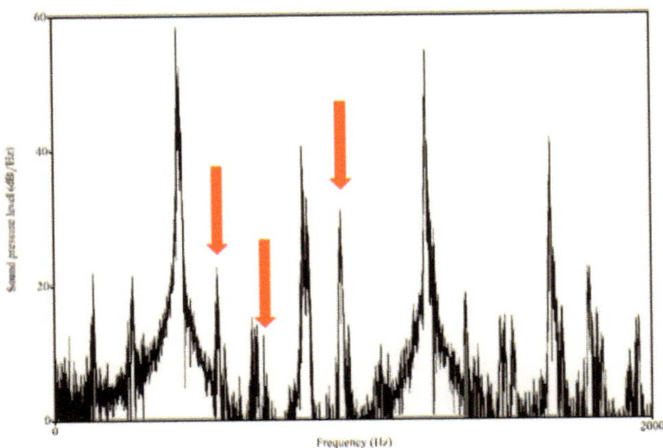

Figure 8B: Power spectrum of the sound displayed in figure 8A in the frequency-range 0-2.000Hz. The red arrows mark the signals of the Complex tones at 546Hz (F1: 130Hz/ F2: 415Hz); 701Hz and 962Hz (F1: 130Hz/ 1st harmonic F2: 832Hz). See text for more details.

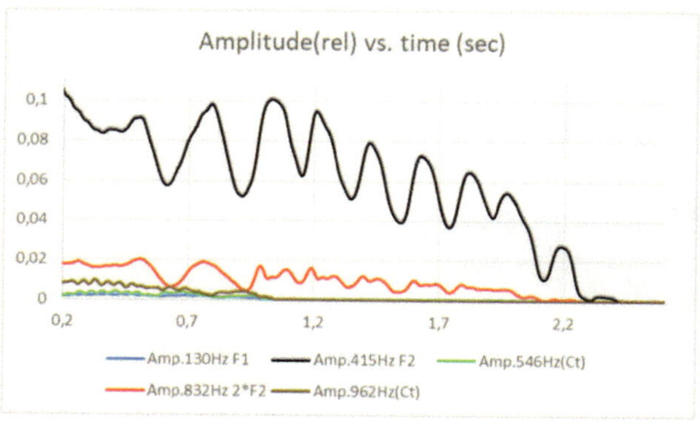

Figure 8C: Amplitude values (rel.) vs. time (sec) of various signals extracted from the power spectrum shown in figure 8B. According to the nomenclature of multiphonic sounds (Ref. 3) the amplitudes of the basic frequencies F1 (130Hz) and F2 (415Hz) are shown as well as the 1st harmonic of F2 (832Hz, 2*F2) and of the complex tones at 546Hz (F1+F2) and at 962Hz (F1+2*F2).

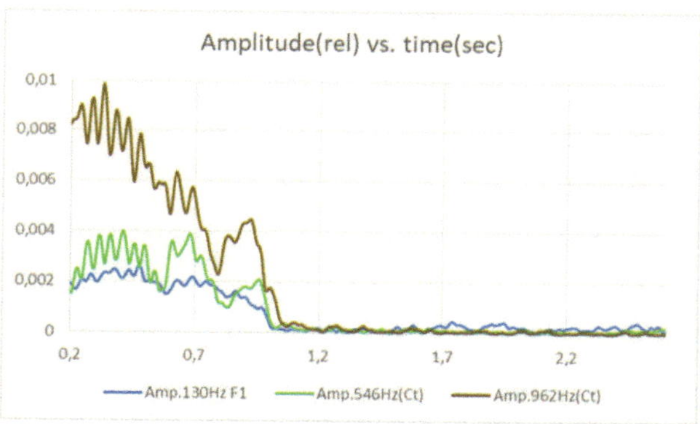

Figure 8D: Amplitude-values (rel.) vs. time (sec) from figures 8C of the signals at 130Hz, 546Hz and 962Hz. The y-axis has been enlarged vs. figure 8C for a better visibility of the data.

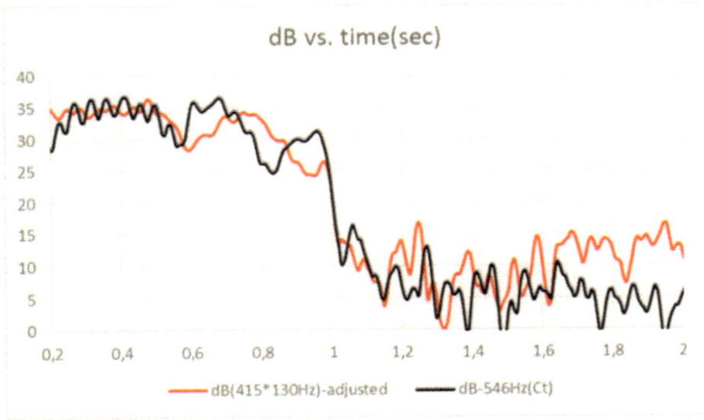

Figure 8E: dB-kinetics of the Complex tone (F1+F2) at 546Hz (black curve) and the calculated and adjusted dB-values of the multiplication of the amplitudes of the oscillations of F1 and F2 (red curve). (Adjustment for same intensity at 0,2-0,4sec)

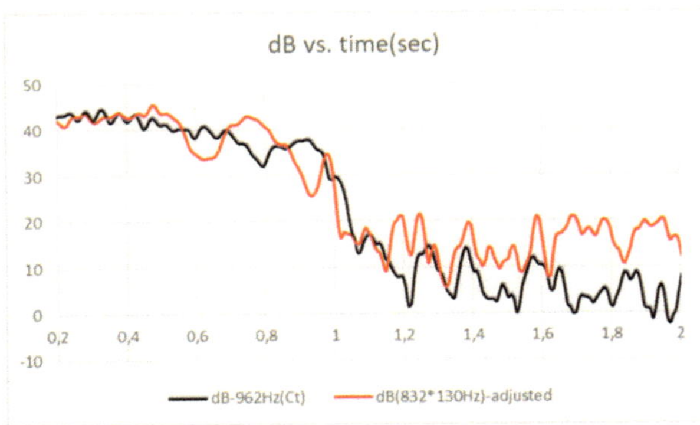

Figure 8F: dB-kinetics of the Complex tone (F1+2*F2) at 962Hz (black curve) and the calculated and adjusted dB-values of the multiplication of the amplitudes of the oscillations of F1 and 2*F2 (red curve). (Adjustment for same intensity at 0,2-0,4sec)

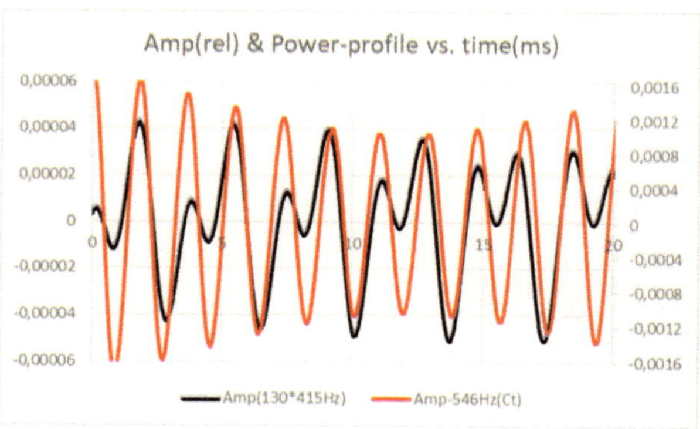

Figure 9A: Amplitude-values of the Complex tone with the frequency 546Hz in a period of 20ms (red curve / related y-axis on the right) and values of the power-profile for F1 (130Hz) and F2 (415Hz) in the same period (black curve / related y-axis on the left).

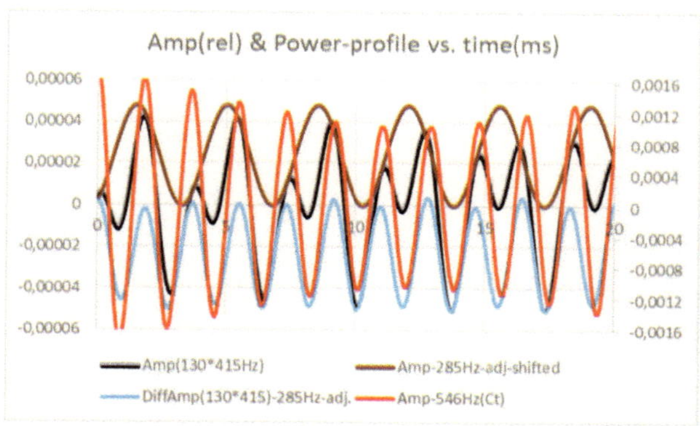

Figure 9B: Red and black curves are identical to the curves in figure 9A, representing the Complex tone at 546Hz (red) and the power-profile (black) of F1 (130Hz) and F2 (415Hz). The brown curve is an artificially generated oscillation with a frequency of 285Hz adjusted in phase and intensity in such a way so that the difference of the power-profile (black curve) and this curve (brown) results in a stable sine-like oscillation with the frequency of 546Hz (blue curve).

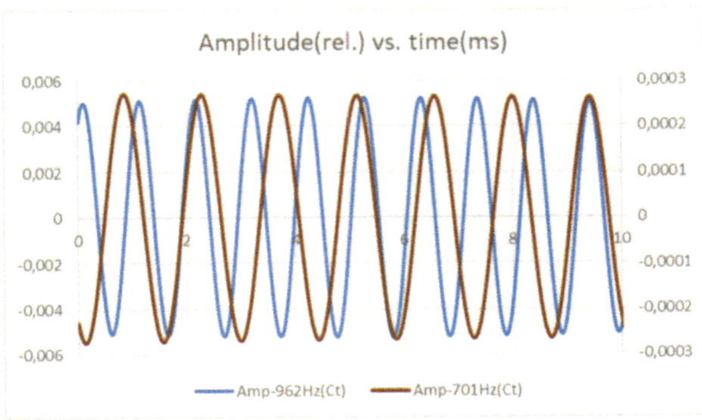

Figure 10A: Amplitude values of the oscillations of the Complex tones at 701Hz (brown curve / related y-axis on the right) and 962Hz (blue curve / related y-axis on the left) extracted from the power spectrum in figure 8B (time period: 10ms).

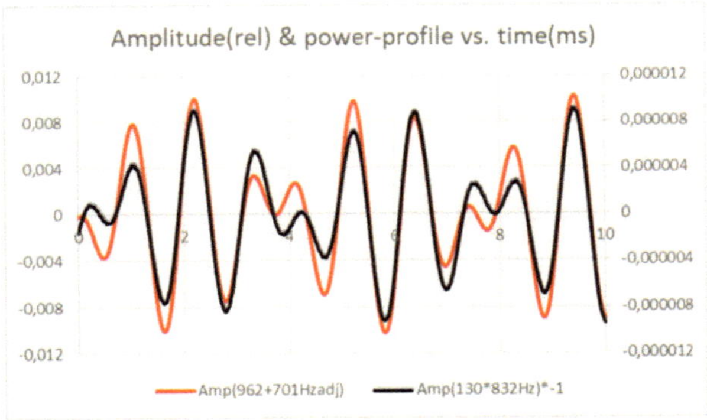

Figure 10B: Comparison of the power-profile (black curve/ y-axis to the right) of the amplitude-values of the oscillations at 130Hz (F1) and 832Hz (2*F2; 1st harmonic of F2) and the sum of the amplitude-values of the Complex tone at 962Hz (2*F2+F1) and at 701Hz (2*F2-F1)(red curve /y-axis to the left). The slight deviation of both curves is caused by a shift of the phase of the Complex tone at 701Hz by 0.12ms (8% of the wavelength) vs. the phase of the corresponding oscillation of the power-profile.

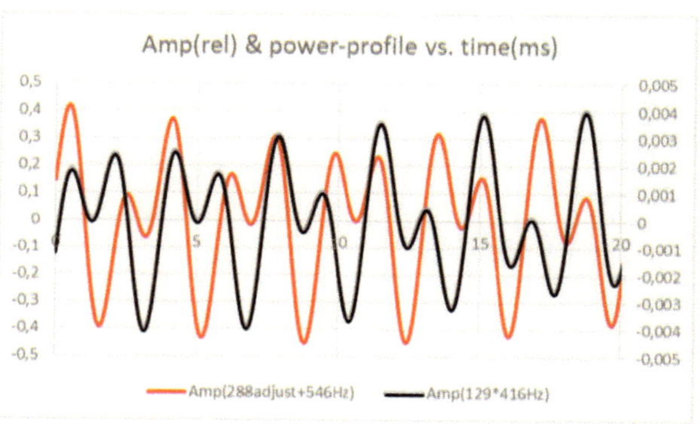

Figure 11A: Data of an analyzed multiphonic sound (downloaded from Ref. 14) generated with a tenor saxophone with basic frequencies at 129Hz (F1) and 416Hz (F2). The red curve (y-axis to the left) shows the sum of the amplitudes-values of the oscillations of the complex tones at 546Hz (F1+F2) and 288Hz (F2-F1; adjusted to the same maximum amplitude as the oscillation at 546Hz). The black curve (y-axis to the right) shows the power-profile of F1 and F2. Time-period: 20ms.

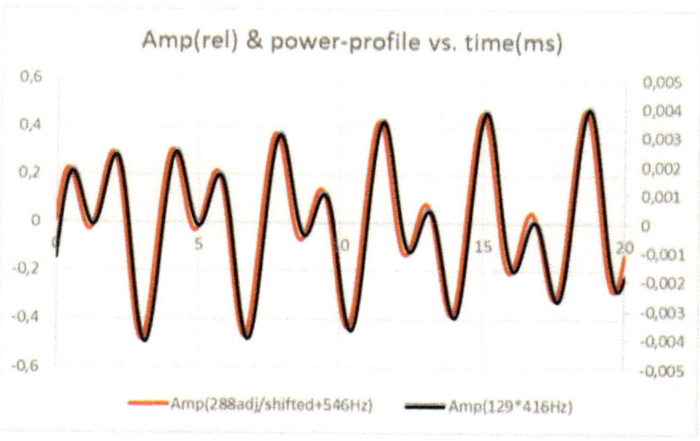

Figure 11B: The black curve (power-profile of F1 and F2; y-axis to the right) is the same curve as in figure 11A. The red curve (y-axis to the left) is a sum of the amplitudes-values of the oscillations of the complex tone at 546Hz (F1+F2) and the complex tone at 288Hz (F2-F1; adjusted to the same maximum amplitude as the oscillation at 546Hz) shifted in time (=phase-shift) by 1,066ms which corresponds to 31% of the wavelength of 288Hz.

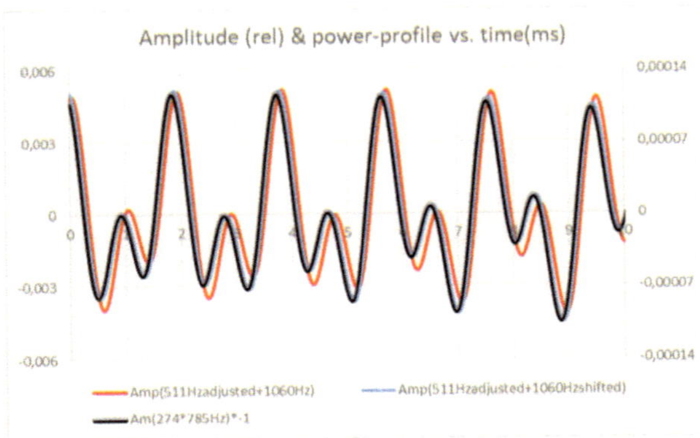

Figure 12: Data of an analyzed multiphonic clarinet sound (Ref. 15; Sound No.60) with the two basic frequencies at 247Hz (F1) and 785Hz (F2). The black curve represents the power-profile of the amplitude-values of F1 and F2 (y-axis to the right) in a time period of 10ms. The red curve (y-axis to the left) is a sum of the amplitude-values of the oscillations of the complex tone at 1060Hz (F1+F2) and at 511Hz (F2-F1; adjusted to the same maximum amplitude as the oscillation at 1060Hz) in the same time period. The blue curve (y-axis to the left) represents the sum of the amplitude-values as in the red curve but with a shifted phase of the oscillation at 1060Hz by 0.068ms which represents 7% of the wavelength. Remark: blue curve is hardly visible as it matches the black curve!

Discussion

Although wine glasses and musical wind instruments are neither comparable in structure nor in material, both have a tube-like cavity in common in which periodic air oscillations are generating an audible sound. The finding that it is possible to generate multiphonic sounds with their typical complex tones as derivatives of two base-frequencies in both "instruments" (Ref. 4-7, 11) can lead to the conclusion that the mechanism to produce those complex tones is of the same nature in wine glasses and wind instruments. A previous study has described complex tones as part of multiphonic sounds generated through a hit-excitation of wine glasses (Ref. 11) – this could be confirmed in this research (see figures 1B, 1C). As the Eigenfrequenzen (natural frequencies) of wine glasses show a typical decay kinetic, it was possible to compare those with the decay kinetics of the complex tones, which are supposed to be derivatives of two respective Eigenfrequenzen (see figures 1 and 2). In some wine glasses we could find a linear correlation of the decay-kinetic and the frequency of the Eigenfrequenzen (see figures 1F; 2D) but neither the complex tones nor the 1st harmonic of the 1st Eigenfrequenz fit into this correlation. It can be concluded that although the 1st harmonic and the

26

complex tones are derivatives of the base frequencies (which are Eigenfrequenzen), those oscillations differ from oscillations of the Eigenfrequenzen either by a stronger damping or a different dynamic of the driving force. The data presented in figures 1, 2 and 5 are examples of the data generated with several wine glasses, which demonstrate that the decays of the amplitudes of the complex tones correlate with the result of the mathematical multiplication of the amplitude-values of the respective Eigenfrequenzen. That means that for the decay of the amplitude of a complex tone, the following equations can be written:

AmpCt(F2+F1)(t) = Amp(F2)(t) * Amp(F1)(t) * E-factor(F2+F1)

AmpCt(F2-F1)(t) = Amp(F2)(t) * Amp(F1)(t) * E-factor(F2-F1)   / for F2>F1

AmpCt(F2+F1)(t) and AmpCt(F2-F1)(t) are the Amplitude values of the complex tones with the frequencies in Hz of "F1+F2" and "F2-F1" (with F2>F1) at a given time (t)

Amp(F2)(t) and Amp(F1)(t) are the Amplitude values of the Eigenfrequenzen F2 and F1 at a given time (t).

The E-factor is a **Constant** named "Efficiency-factor" as it defines the ratio of the values for AmpCt(F2+F1)(t) and AmpCt(F2-F1)(t) vs. the result of the multiplication: Amp(F2)(t)*Amp(F1)(t).

In the introduction-part, the "power-profile" was defined as multiplication of the amplitude-values of two sine-like oscillations of different frequencies, so we can interpret the E-factor as a parameter describing how efficient the "power-profile" will generate the complex tones. Our analysis of the decay-kinetics delivered E-factor values in the range of 0,1 - 14,6 for various wine glasses and complex tones. So it can be assumed that each wine glass has a specific E-factor for each of the complex tones which the wine glass can generate. Often no complex tones with frequencies "F2-F1" (with F2<F1) could be detected, which means that the E-factor for these frequencies were very small or even zero. If the Amplitude-values had the unit Pascal (Pa) the E-factor would have the unit (Pa^{-1}) according to the equation above.

Based on these findings, the following hypothesis for the generation of complex tones in wine glasses has been developed:

a) The "power-profile" of two sine-like oscillations with different frequencies generated in parallel through the oscillations of the rim and bowl of the wine glass is the driving force for the Complex tones.

b) The "power-profile" of the oscillations "F1" and "F2" is acting as a force against those parts of the bowl and rim of the wine glass which are more or less fixed (= "no-areas"; see text above) and therefore cannot follow the oscillation of "F1" and "F2" – see for details figure 4.

c) At those areas which are fixed for the oscillations of "F1" and "F2" the power-profile will have a maximum effect and may force these parts of the glass to oscillate with the frequencies of the power-profile which are: "F2-F1" and "F2+F1".

d) The power-profile is acting as a force to generate the oscillations of the complex tone from the inside of the bowl against the walls of the bowl of the wine glass. (Remark: the volume inside the bowl of a wine-glass could also be interpreted as a cavity or short tube – a similarity to wind instruments, as the tube of those instrument is the area where the standing waves are generated)

Following this hypothesis, we would expect, that the phases of the oscillations of the complex tones with the frequencies "F2-F1" and "F2+F1" should match the phases of the two oscillations of the power-profile of the Eigenfrequenzen "F1" and F2.

With an error of <5% we found the following two phenomena (see also figures 6 & 7):

1) The phases of both complex tones match the phases of both oscillations of the power-profile.

2) The phase of one complex tone matches the phase of the respective oscillation of the power-profile, whereas the other complex-tone matches the inverted respective oscillation of the power-profile.

These findings are strong arguments for the hypothesis mentioned above. And if we assume that the bowl of the wine glass and the tube of a wind instrument may have a similar meaning for the generation of complex tones within multiphonic sounds, we should be able to find the same or similar effects with wind instruments as those described above for wine glasses.

In fact, such similar effects have been presented in this study for multiphonic sounds of tenor saxophones (see figures 8-11 / Ref. 13, 14) and of a clarinet multiphonic (see figure 12 / Ref. 15). Analyzed data from other multiphonic sounds (not shown in this publication) from the same sources (Ref. 13-15) confirm the presented findings. There is one inconsistency in the data from wind instruments compared to wine glasses: For wine glasses the phases of the oscillations of the complex tones "F2-F1" and "F2+F1" and of the two oscillations within the power-profiles of "F2" and "F1" are identical or inverted whereas in wind instruments (tenor saxophone & clarinet examined) a shift of the phase of one of the two complex tones by >5% but <45% of the corresponding wavelength is necessary to achieve a match with the respective oscillation of the power-profile. This discrepancy cannot be explained yet. But it is likely that if the mechanism to produce complex tones is the same in wine glasses and wind instruments, the more complex structure and especially the long tube or cone of a wind instrument compared to the rather short bowl of a wine glass might be responsible for this discrepancy. It would be interesting to see what effects on the complex tones could be identified in less complex flute-like instruments described by Vergez. et al. (Ref. 10). But also further analysis of multiphonic sounds of clarinets and saxophones might deliver some information to understand the observed phenomena as the pressure values resulting from standing waves within the tube of the clarinet (Ref. 21) and the cone of the saxophone (Ref. 22) as well as the shapes of both instruments differ significantly. It would also be of interest whether complex tones of multiphonic sounds of the trombone (Ref. 23) show similar effects although the technique of the player to produce those sounds on the trombone (Ref. 24) differ from those of the clarinet or saxophone. The proposal of Linke et al (Ref. 9) that the formation of an impulse pattern within the instrument is the key driver to generate a multiphonic sound is in line with the idea of the power-profile being the driving force for the complex tones as part of the multiphonic sound. The power-profile can be understood as a quasi-periodic impulse, generating periodic oscillations of certain areas of the walls of the cone (e.g. saxophone) or tube (e.g. clarinet) or the bowl of a wine glass. The observed different intensities of the complex tones could be caused by the following factors: a) Intensities of the two basic frequencies "F1" and "F2", b) Adequate areas of the cone or tube of the instrument which are sensitive to the impulses of the power-profile built by "F1" and "F2" and c) Ability of these "sensitive areas" to oscillate at the respective frequencies "F2+F1" and "F2-F1" (= resonance ability) which are transmitted through the impulse of the power-profile. The combined effects of factors b) and c) might be expressed by the above defined Efficiency-factor (E-factor).

Further studies on multiphonic sounds generated by different wind instruments might be helpful to find further arguments for the hypothesis presented in this study. Especially wind instruments with a less complex structure as clarinets or saxophones might be ideal candidates for these investigations and might help to explain the observed phase shift of one of the complex tones vs. the power-profile observed for clarinets and saxophones.

References:

1) M. Proscia, P. Riera, M. Eguia; "Comparative study of saxophone multiphonic tones. A possible perceptual categorization"; Proceedings 12th international conference on Music Perception and Cognition; July 2012 Thessaloniki; pp 822

2) M. Proscia, P. Riera, M. Eguia; "A timbral and musical performance analysis of saxophone multiphonics morphing"; Procceedings international symposium on musical acoustics; June 2017 Montreal; pp 9-12

3) P. Riera, M. Proscia, M. Eguia; "A comparative study of saxophone multiphonics: Musical, psychophysical and spectral analysis"; Journal of new Music research; 2014; DOI: 10.1080/09298215.2013.860993

4) S.Watts; "Spectral immersions: A comprehensive guide to the theory and practice of bass clarinet multiphonics"; Thesis; Keele University UK; 2015

5) J.Backus, "Multiphonic tones in the woodwind instruments." The Journal of the Acoustical Society of America, 63(2), pp 591–599; 1978.

6) A. Benade; „Fundamentals of musical acoustics"; Second revised edition, Dover publications 1990, ISBN: 139780486264844

7) A.Rehm; J.Bullenkamp; "Analysis of spectral parameters of mulitphonic sounds of string- and wind-instruments and introduction of a mathematical model to simulate the frequencies and intensities of the harmonics and the complex tones of multiphonics generated with woodwinds;" ISBN: 97833466652799; 2022; Deutsche Nationalbibliothek; http://dnp.d-nb.de

8) D. Keefe, D. Laden; "Chaotic dynamics of woodwind multiphonics"; Journal Acoustic Society America Suppl.1, Vol.86, 118th meeting; 1989

9) S.Linke, R.Bader, R. Mores; „Multiphonic modeling using Impulse Pattern Formation (IPF)";preprint: arXiv: 2201.05452v1; January 2022

10) S.Terrien, C.Vergez, P.Cuadra, B.Fabre; "Experimental analysis of non-periodic sound regimes in flute-like musical instruments"; JASA 149 (3) pp 2100-2108; March 2021

11) A.Rehm; "Wine glasses as multiphonic instruments forced to sound like monophonic instruments!"; ISBN: 9783346838889; 2023; Deutsche Nationalbibliothek; http://dnp.d-nb.de

12) References and details about software "Praat" see website: https://www.fon.hum.uva.nl/praat/ (status 1/2022)

13) Website: Worldwide Sax; https://worldwidesax.com/the-vault/education/sound-files/; soundfiles downloaded in 2/2023

14) Website: https://www.baerenreiter.com/materialien/weiss_netti/saxophon/mehrklang-auswahl.htm; Autoren: M. Weiss, G. Netti; downloaded March 2021

15) Website: https://heatherroche.net/2018/09/13/27-easy-bb-clarinet-multiphonics/; download of Clarinet Multiphonics September 2020

16) G.Denninger; „Das Ohr trinkt mit"; Physik in unserer Zeit; Physik in unserer Zeit; Vol. 44; pp 142-146; 2013; DOI: 10.1002/piuz.201301327

17) Cosmol software-App of the Technical University Munich (TUM), website: https://apps.vib.ed.tum.de:2037/comsol-software-license-agreement; date of usage: 2/2023

18) L.Mäder, L.Moheit, L.Kastenhuber and S.Marburg; „Zur Akustik von Weingläsern"; Akustik Journal 03/18, pp 7-14; 2018

19) W.Poomarin and I.Jacobs; "Wine glass oscillations"; website: https://kvis.ac.th/userfiles/files/Wine%20glass%20oscillations.pdf; download 2/2023.

20) S.Yang; "Wine glass acoustics"; website: https://studylib.net/doc/18265335/wine-glass-acoustics-how-does-the-frequency-vary-with-the; download: 2/2023

21) A.Rehm, D.Gaebel, T.Lakatos, C.Valk, S.Weber; "Schallwellenanalyse des Sounds professioneller SaxophonspielerInnen Teil6. Vorstellung eines einfachen Modells zur Reed-Bewegung während einer stabilen Tonerzeugung und Auswirkung der Reed-Bewegung auf den Saxophon-Sound"; ISBN: 9783668975118; 2019; Deutsche Nationalbibliothek; http://dnp.d-nb.de

22) D.Lucchetta, L.Schiaroli, G.Battista, M.Martarelli, P.Castellini; "Experimental acoustic modal analysis of a tenor saxophone"; JASA 152 (2) pp2629-2640; November 2022

23) Website: https://www.youtube.com/watch?v=r9lrjUatNA8; download of Trombone Multiphonics; December 2021

24) L. Velut, C.Vergez, J.Gilbert; „Measurements and time-domaine simulations of multiphonics in the trombone"; Acoustical Society of America; 2016; pp 2876-2887